habitat explorer

City Explorer

Neil Morris

Raintree

www.raintreepublishers.co.uk
Visit our website to find out more information about **Raintree** books.

To order:
 Phone 44 (0) 1865 888112
 Send a fax to 44 (0) 1865 314091
Visit the Raintree Bookshop at **www.raintreepublishers.co.uk** to browse our catalogue and order online.

First published in Great Britain by Raintree, Halley Court, Jordan Hill, Oxford OX2 8EJ, part of Harcourt Education.
Raintree is a registered trademark of Harcourt Education Ltd.

© Harcourt Education Ltd 2004
First published in paperback in 2005
The moral right of the proprietor has been asserted.

Editorial: Nick Hunter and Diyan Leake
Design: Michelle Lisseter
Picture Research: Maria Joannou
Production: Jonathan Smith

Originated by Repro Multi Warna
Printed in China by WKT Company Limited

ISBN 1 844 43457 5 (hardback)
08 07 06 05 04
10 9 8 7 6 5 4 3 2 1

ISBN 1 844 43465 6 (paperback)
09 08 07 06 05
10 9 8 7 6 5 4 3 2 1

British Library Cataloguing in Publication Data
Morris, Neil
City Explorer
577.5'6
A full catalogue record for this book is available from the British Library.

Acknowledgements
The publishers would like to thank the following for permission to reproduce photographs: Alvey & Towers p. **10**; Ardea p. **6**; Bruce Coleman Collection pp. **5** (Orion Press), **20** (Jane Burton), **27** (Kim Taylor), **28** (Jane Burton); Corbis pp. **4**, **12**, **22**; Ecoscene p. **18** (Martin Jones); FLPA pp. **9**; (Alwyn J. Roberts), **13**, **17**, **24** (E & D Hosking); Naturepl.com pp. **14** /(William Osborn), **25** (Warwick Sloss); NHPA pp. **8** (David Woodfall), **21** (Stephen Dalton), **23** (Stephen Dalton), **26**; Oxford Scientific Films pp. **7** (David Boag), **15** (Norbert Rosing), **16** (Caroline Aitzetmuller), **19** (Mark Hamblin); Photographers direct.com p. **11**; Photolibrary.com p. **29** (John Tuffin)

Cover photograph of a peregrine falcon reproduced with permission of RSPCA Photolibrary (Martin Dohrn)

The publishers would like to thank Louise Spilsbury for her assistance in the preparation of this book.

Every effort has been made to contact copyright holders of any material reproduced in this book. Any omissions will be rectified in subsequent printings if notice is given to the publishers.

Disclaimer
Third-party website addresses are provided in good faith and for information only. The publishers disclaim any responsibility for the material contained therein. All the Internet addresses (URLs) given in this book were valid at the time of going to press. However, due to the dynamic nature of the Internet, some addresses may have changed, or sites may have ceased to exist since publication. While the author and publishers regret any inconvenience this may cause readers, no responsibility for any such changes can be accepted by either the author or the publishers.

The paper used to print this book comes from sustainable resources.

Contents

Any words appearing in the main text in bold, **like this**, are explained in the Glossary.

Arriving in the city

Imagine you are sitting on a train, travelling into a big city. Just half an hour ago you were speeding through the countryside and saw fields as you looked out of the window. But now the view is changing. You see more and more houses and factories, until finally you pull into a large station.

Sights and sounds

As you walk outside, the first thing you hear is the loud noise of traffic. When you look up, you see that most of the buildings around you are very tall. They seem to be reaching right up into the sky.

Cities can have very busy traffic and tall buildings.

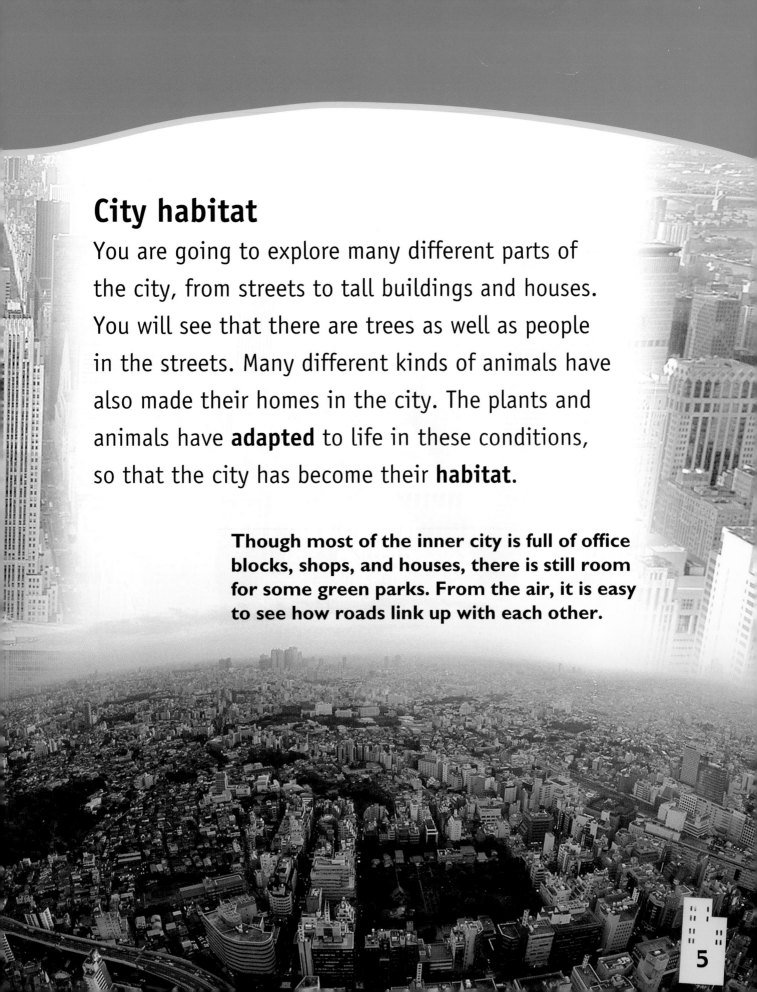

City habitat

You are going to explore many different parts of the city, from streets to tall buildings and houses. You will see that there are trees as well as people in the streets. Many different kinds of animals have also made their homes in the city. The plants and animals have **adapted** to life in these conditions, so that the city has become their **habitat**.

Though most of the inner city is full of office blocks, shops, and houses, there is still room for some green parks. From the air, it is easy to see how roads link up with each other.

Concrete cliffs

Some birds have **adapted** to life on top of tall city buildings. Herring gulls nest there. They are not always easy to see, because they are **camouflaged** against colours of the buildings. Parent gulls bringing up chicks often visit the nearest rubbish dump to find food. While they are away, the chicks are in danger from kestrels. These **predators** have also taken to nesting on high ledges, and they usually hunt over parks.

A herring gull protects its chick, which will be brought up in the city.

Pigeons like cities because they can find food there all year round.

The animals and plants that we find in the city have all adapted to life there. This means that they have changed to better suit the conditions in which they live. They have features that allow them to survive in their **habitat**.

Pigeons

The most common bird you will see in a city is the pigeon. In a city square, you might see hundreds of them gathered together. Pigeons can live well off scraps of food they find in the street. They often use plastic drinking straws and lengths of wire to build their flimsy nests.

Explorer's notes

Some common birds on city buildings:
- pigeon
- herring gull
- kestrel.

In the street

As you walk through
the city streets, you will
see that many different trees
grow there. Though they often
have very little soil at the base
of their trunk, the trees' long roots
reach down beneath pavements and roads
to find water. The **bark** of the plane tree
flakes off regularly. This means that the
plane tree's trunk does not become coated
with dirt and air **pollution**.

**London plane trees can
grow up to 35 metres high.**

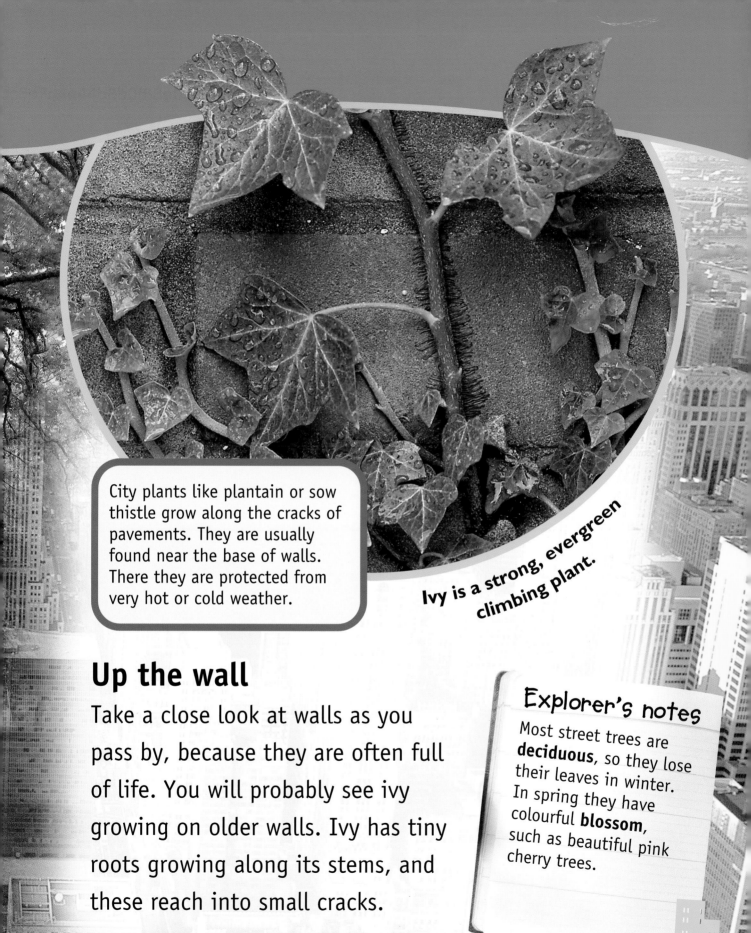

City plants like plantain or sow thistle grow along the cracks of pavements. They are usually found near the base of walls. There they are protected from very hot or cold weather.

Ivy is a strong, evergreen climbing plant.

Up the wall

Take a close look at walls as you pass by, because they are often full of life. You will probably see ivy growing on older walls. Ivy has tiny roots growing along its stems, and these reach into small cracks.

Explorer's notes

Most street trees are **deciduous**, so they lose their leaves in winter. In spring they have colourful **blossom**, such as beautiful pink cherry trees.

Along the railway

From a bridge over the railway, you have a good view of an **embankment**. This is left undisturbed most of the time – apart from trains regularly whizzing past! This means that the embankments provide a safe haven for many different plants. These include broom, ragwort, cow parsley and buddleia.

Railway embankments are like wild gardens in the city.

Animal homes

Railway embankments can make a good home for large **mammals**. Badgers sometimes dig their **setts** and foxes often dig their **earths** there. These animals' burrows are quite large, and their entrances are usually about the size of a football. If you can see smaller holes, they probably belong to rabbits.

Foxes often make their home along railway embankments.

Just as people use the railways to travel into cities, animals do too. Embankments are like green corridors, which foxes and rabbits use to move towards the city centre. Foxes have even been found at Waterloo station, in central London.

By the river

The river is a good place to find and watch wildlife. You can explore the riverbank and the water by walking along the **towpath**. You will probably see lots of ducks and geese, and you may be lucky enough to see swans. Many of the waterbirds are **migratory**, which means that they make long journeys with the changing seasons.

There may be swans on rivers in cities.

Cities began as small settlements many years ago, and almost all grew up beside a river. This was a good source of fresh water for the people of the growing village and town.

Fishing methods

If you see a grey heron beside the riverbank, stand very still and watch it from a distance. Herons do not like being disturbed when they are fishing. They stand beside the water, patiently waiting and watching for passing fish. When it sees one within range, a heron will stab its **prey** with its long pointed beak.

This grey heron is fishing on a canal bank.

Explorer's notes

Water **habitats** in the city:
- rivers
- canals (manmade rivers)
- reservoirs (manmade lakes where water is stored)
- docks
- lakes and ponds (in parks and gardens).

Waste and rubbish

There is always the odd patch of waste ground in the city, but it is never empty for long. Seeds are blown by the wind from nearby plants. They start to grow on the wasteland. A gardener might call these plants weeds, but an explorer is more likely to see them as interesting wild plants. Thistles grow well on waste ground. The thistle is a beautiful plant but prickly if you happen to touch one.

Thistles are favourite feeding places for goldfinches, which peck out and eat the seeds.

If you watch a patch of overgrown wasteland carefully, you might see some of the plant stems move as shrews and mice dash about.

Gulls search a rubbish tip for food.

Scraps and waste

Many birds find rubbish tips a good source of food, especially during winter. You will see gulls searching for scraps. Other common birds that **scavenge** are crows and starlings. You will not be able to see them, but many small **insects** get into the waste and help break it down. These include houseflies, bluebottles, midges and springtails.

Park life

Parks provide a haven for wildlife. One of the first things you will notice in any city park is the wonderful variety of trees. These vary throughout the year. In spring, many are covered in beautiful **blossom**, and then in autumn they bear fruit.

A horse chestnut tree in bloom brightens up any park.

Tree blossom is a mass of flowers. The colour and scent of the flowers attract **insects**, such as bees, which feed on a sugary liquid called **nectar**.

Mallards go bottom-up to feed in the pond.

Variety show

There is a great variety of animal life in city parks. Grey squirrels leap through the branches of trees, coming to the ground to bury or find their hidden nuts.

On the park pond you might see mallards and other ducks. Look out for frogs or their spawn. This floats on the pond and looks like a mass of round balls of jelly. These are the frogs' eggs, which after about two weeks hatch into tiny wriggling tadpoles.

Explorer's notes

Trees and their fruit
(which contain seeds):
- chestnut – sweet chestnuts
- horse chestnut – conkers
- oak – acorns
- pine – cones
- sycamore – winged seeds

In the churchyard

There are plenty of churches and churchyards in the city. Apart from the digging of new graves, some churchyards have been left undisturbed and unchanged for hundreds of years. This means that they are like ancient meadows, with flowering plants such as cowslips, ox-eye daisies and knapweed. You will probably find greyish green **lichen** growing on old gravestones.

Many plants grow in city churchyards.

Yew trees are traditionally found in churchyards. They make good nesting trees for goldcrests and other birds. Be careful when exploring yews: their **bark**, leaves and seeds are all poisonous.

Bats in the belfry

Bats like to live in an enclosed space. This makes the belfry on top of a church tower the ideal home for them. Watch out at sunset, when they leave the belfry to hunt **insects**.

Hedgehogs usually come out at night to hunt for slugs, snails and earthworms.

Living house

Imagine you have been invited by friends to end your city exploration by spending some time at their house. They might think you would see a few birds in their garden, but they are probably unaware of the amount of wildlife inside their house. There is usually very little plant life, except for a few pot plants, because of the lack of water and soil.

A house spider checks its web.

House spiders spin webs shaped like hammocks. They are usually in dark corners, but they are not there for resting in – they are to catch the spiders' **insect prey**, mainly houseflies.

A house martin feeds its chicks in the nest.

Under the eaves

House martins build their mud nests against a wall or under the **eaves**. There may even be lots of nests crowded together, especially if the house is near wet ground or a stream so that there is plenty of nest-building material. Swifts usually prefer to nest inside roof spaces, where they might face some competition. Bats also like to roost there, hanging upside down from the rafters.

Wild garden

Gardeners usually like to have control of what they grow in their gardens. Because of this, they often call wild plants weeds. This is because they are growing where they are not wanted, so gardeners have to do a lot of weeding. One of the most common plants they look for is the dandelion. It spreads very successfully. Its seeds are blown away by the wind and float to another piece of soil.

Bindweed is a plant that grows up walls and other plants.

These butterflies are feeding on buddleia.

Attracting butterflies

The buddleia is so popular with butterflies that we often call it the 'butterfly bush'. This plant has masses of tiny flowers on long spikes. The flowers have a strong scent. The combination of colour and smell attracts butterflies, which come to feed on the flowers' **nectar**.

In some parts of the world, you might come across poisonous spiders in the garden. In Australia, funnel-web spiders live in burrows under rocks or leaves. Their bite is painful and can be very dangerous to humans.

Explorer's notes

Insects to look out for in the garden:
- butterflies (peacock, tortoiseshell, red admiral)
- bees, wasps, ladybirds, flies.

Spiders are arachnids, not insects (with eight legs instead of insects' six).

Dawn chorus

People who live in cities are often woken first thing in the morning by the beautiful sound of birds singing. Just before it gets light you can hear this dawn chorus. Each bird's song is made up of a series of notes in a set pattern, repeated over and over again. Male birds sing to attract a mate and to warn other males to stay out of their **territory**.

This blackbird has chosen a television aerial for its singing perch.

Special songs

Very often the first song in the dawn chorus is that of a blackbird, with a rich sound like a flute. One of the easiest birds to recognize is the wood pigeon, which makes a call that sounds like 'woo-coooo-coo, coo-coo'.

During the day you will probably see many different kinds of birds in the city garden. Digging gardeners are often joined by a red-breasted robin, which might sit on a nearby fence and visit the disturbed earth looking for an easy meal of worms.

Robins know that digging disturbs earthworms.

Life under ground

There is lots of underground life in city gardens and parks. One of the largest underground animals is the mole. Moles dig tunnels as they look for food – mainly earthworms and **insects**. You might find some molehills, which are made up of the waste soil from a mole's network of tunnels. Their **prey**, earthworms, also burrow through underground tunnels, but much smaller ones. They eat their way through the soil, feeding on dead plant material.

A mole emerges from its tunnel. This is a rare sight.

Insects

Many insects have an underground life, too. Female mining bees make tiny tunnels in the ground, where they lay their eggs. Many insect larvae, such as leatherjackets (the young stage of craneflies) feed on roots in the soil.

Black ants live under paving slabs or large stones in the garden. One day in summer, when the weather is warm, winged ants suddenly swarm out of the nest. The winged ants are the many males and few queens, and they take to the air to mate.

Explorer's notes

Most insects change as they go through four life stages: egg, larva, pupa and adult. A caterpillar is the larva of a butterfly.

Winged black ants leave their nest.

Changing cities

The world's towns and cities are growing. This means that they are also a fast-spreading **habitat** for wildlife. Many animals have **adapted** so well to **urban** life that people regard them as **pests.** This has probably always been true of mice and rats, and it is increasingly the case with foxes and grey squirrels. Foxes **scavenge** in the streets, tearing open rubbish bags, and squirrels move into lofts if they can.

This red fox has found plenty of food scraps.

Cleaning up the environment

In recent years, people have tried to clean up their cities. In some cases, this is also very helpful for wildlife. One of the best examples is the cleaning-up of rivers. By stopping factories dumping waste into rivers, we have also attracted water plants, fish, birds and **insects**.

This is Hyde Park in Sydney, Australia, which is very different from Hyde Park in London. They are both good to explore.

Many of the world's biggest cities are famous for their parks. Hyde Park, in London (UK), has a large lake called the Serpentine. Ueno Park, in Tokyo (Japan), is well known for its beautiful spring displays of cherry **blossom**.

You may live in a city or have the chance to visit. If you do, look out for different types of **habitats**, and the different types of plants and animals that live there. Perhaps you could compare different birds in a garden or a nearby park.

Using the Internet

Explore the Internet to find out more about city habitats. Websites can change, but if some of the links below no longer work, don't worry. Use a kid-friendly search engine, such as Yahooligans, and type in keywords such as 'garden', 'frogs' or the name of an area like a river or a park.

Websites

www.fund.org and www.urbanwildlifesociety.org
The Fund for Animals and the Urban Wildlife Society aim to help people solve problems of living with **urban** wildlife.
www.garden-birds.co.uk
Information about all kinds of garden birds (and recordings of their songs).
www.mammal.org.uk
The Mammal Society works to protect **mammals** and halt the decline of threatened species.
www.plantlife.org.uk
Plantlife, the wild-plant conservation charity, specializes in plant protection.

Glossary

adapt   change in order to suit the conditions

bark   outer covering of a tree trunk

blossom   mass of small flowers on a tree

camouflage   blending in with the surroundings, for example by changing colour

deciduous   a word describing trees that shed their leaves in autumn

earth   fox's underground burrow

eaves   part of a roof that hangs over a wall

embankment   raised slope beside a railway line

habitat   natural environment, or home, of an animal or a plant

insect   small animal with six legs

lichen   plant-like growth found on rocks or trees

mammal   warm-blooded animal that feeds its young on milk

migratory   a word referring to animals that go on a long journey, usually with the changing seasons

nectar   sweet liquid produced by flowering plants

pest   animal that causes damage

pollution   damage by harmful substances

predator   animal that hunts and kills other animals for food

prey   animal that is hunted by another animal for food

scavenge   search through rubbish for food

sett   badger's underground burrow

territory   area that an animal considers as its own

towpath   path beside a river or canal

urban   relating to a town or city (as opposed to the countryside)

Index